Comprehensive Summaries of Uppsala Dissertations
from the Faculty of Science and Technology 456

Minuscular Sculpturing

BY

GREGER THORNELL

ACTA UNIVERSITATIS UPSALIENSIS
UPPSALA 1999

Dissertation for the Degree of Doctor of Philosophy in Materials Science presented at Uppsala University in 1999

ABSTRACT

Thornell, G. 1999. Minuscular Sculpturing. Acta Universitatis Upsaliensis. *Comprehensive Summaries of Uppsala Dissertations from the Faculty of Science and Technology* 456. 42 pp. Uppsala. ISBN 91-554-4480-6.

Most of the processes employed in microstructure technology have been adopted from the fabrication scheme of integrated circuits (IC). Although being extremely reliable and cost efficient, they suffer from poor material and geometrical flexibility. In addition their spreading is hampered by the need for expensive equipment and certain environmental conditions. In the rapidly expanding field of microelectromechanical systems there is a need for truly three dimensional structures, not feasible by planar IC processes, as well as a wish to use and integrate materials with properties other than those of semiconductors and thin metal films. This treatise deals with recent and completely new process technologies with a potential of bridging the gap between needs and tools in this field, and relates them to conventional microprocessing.

Methods to superimpose, or even induce, a virtual wet etch anisotropy in almost any insulating material by swift heavy ions have been developed. By this, microstructures of arbitrary lateral geometry and inclination have been realised in amorphous materials, like polymers and glasses, as well as in highly ordered materials like quartz.

Two techniques to minutely structure conductive materials have been studied. The subtractive electro-discharge machining was successfully combined with wet etching to improve the surface finish and remove the relatively thick heat affected surface layer of micromechanical elements. With the additive localised electrodeposition relying on the free and direct writing of metal structures in an electrolyte with the help of a microelectrode, resolution and surface finish were improved by adding a complexing agent and supporting electrolyte.

Finally, a method to build structures drop-by-drop from dispersed solutions with particles ejected by a micromachined jet nozzle was developed, and applied to a ceramic slurry of lead zirconate titanate - a piezoelectric ceramic.

Greger Thornell, Materials Science, The Ångström Laboratory, Uppsala University, Box 534, SE-751 21 Uppsala, Sweden

ISSN 1104-232X
ISBN 91-554-4480-6

Printed in Sweden by Eklundshofs Grafiska AB, Uppsala 1999

ACTA UNIVERSITATIS UPSALIENSIS
Comprehensive Summaries of Uppsala Dissertations from the Faculty of Science and Technology 456

Distributor: Uppsala University Library, Box 510, SE-751 20 Uppsala, Sweden

Greger Thornell

Minuscular Sculpturing

Dissertation in Materials Science to be publicly examined in the MIC-Aula, on May 28, 1999, at 1.15 p.m. for the Degree of Doctor of Philosophy. The examination will be conducted in English.

ABSTRACT

Thornell, G. 1999. Minuscular Sculpturing. Acta Universitatis Upsaliensis. *Comprehensive Summaries of Uppsala Dissertations from the Faculty of Science and Technology* 456. 42 pp. Uppsala. ISBN 91-554-4480-6.

Most of the processes employed in microstructure technology have been adopted from the fabrication scheme of integrated circuits (IC). Although being extremely reliable and cost efficient, they suffer from poor material and geometrical flexibility. In addition their spreading is hampered by the need for expensive equipment and certain environmental conditions. In the rapidly expanding field of microelectromechanical systems there is a need for truly three dimensional structures, not feasible by planar IC processes, as well as a wish to use and integrate materials with properties other than those of semiconductors and thin metal films. This treatise deals with recent and completely new process technologies with a potential of bridging the gap between needs and tools in this field, and relates them to conventional microprocessing.

Methods to superimpose, or even induce, a virtual wet etch anisotropy in almost any insulating material by swift heavy ions have been developed. By this, microstructures of arbitrary lateral geometry and inclination have been realised in amorphous materials, like polymers and glasses, as well as in highly ordered materials like quartz.

Two techniques to minutely structure conductive materials have been studied. The subtractive electro-discharge machining was successfully combined with wet etching to improve the surface finish and remove the relatively thick heat affected surface layer of micromechanical elements. With the additive localised electrodeposition relying on the free and direct writing of metal structures in an electrolyte with the help of a microelectrode, resolution and surface finish were improved by adding a complexing agent and supporting electrolyte.

Finally, a method to build structures drop-by-drop from dispersed solutions with particles ejected by a micromachined jet nozzle was developed, and applied to a ceramic slurry of lead zirconate titanate - a piezoelectric ceramic.

Greger Thornell, Department of Materials Science, The Ångström Laboratory, Uppsala University, Box 534, SE-751 21 Uppsala, Sweden

To Dr Johansson's stomach with deep gratitude

Outside of a dog, a book is a man's
best friend. Inside of a dog
it's too dark to read.

/Groucho Marx

LIST OF PAPERS

The thesis comprises the following scientific reports, which will be referred to in the text by their Roman numerals.

I *"Design and Fabrication of a Gripping Tool for Micromanipulation"*
Greger Thornell, Mats Bexell, Jan-Åke Schweitz, and Stefan Johansson
Sensors and Actuators, A53, pp. 428-433 (1996).

II *"Heavy Ion Induced Etch Anisotropy in Single Crystalline Quartz"*
Klas Hjort, Greger Thornell, Reimar Spohr, and Jan-Åke Schweitz
Proc. of IEEE MEMS, San Diego, USA, pp. 267-271 (1996).
(A more archival version of this paper was abused in Appl. Phys. Lett. 69 (22), pp. 3435-3436 (1996).)

III *"Anisotropy-Independent Through Micromachining of Quartz Resonators by Ion Track Etching"*
Greger Thornell, Klas Hjort, Bruno Studer, and Jan-Åke Schweitz
IEEE Transactions on Ultrasonics, Ferroelectrics, and Frequency Control, vol. 44, no. 4, pp. 829-838 (1997).

IV *"Micromachining by Ion Track Lithography"*
Greger Thornell, Reimar Spohr, Elbert Jan van Veldhuizen, and Klas Hjort
Sensors and Actuators, A 73, pp. 176-183 (1999).

V *"X-Cut Miniature Tuning Forks Realised by Ion Track Lithography"*
Greger Thornell, Håkan Rapp, and Klas Hjort
submitted to IEEE Transactions on Ultrasonics, Ferroelectrics, and Frequency Control in November 1998.

VI *"Residual Stress in Sputtered Gold Films on Quartz Measured by the Cantilever Beam Deflection Technique"*
Greger Thornell, Fredric Ericson, Christer Hedlund, Johan Öhrmalm, Jan-Åke Schweitz, and Guy Portnoff
accepted for publication in IEEE Transactions on Ultrasonics, Ferroelectrics, and Frequency Control (scheduled to appear in July 1999).

VII *"High Resolution 3-D Microstructures Made by Localised Electrodeposition of Nickel"*
Anders Jansson, Greger Thornell, and Stefan Johansson
submitted to Journal of the Electrochemical Society in April 1999.

VIII *"Desktop Microfabrication - Initial Experiments with a Piezoceramic"*
Greger Thornell, Lena Klintberg, Thomas Laurell, Johan Nilsson, and Stefan Johansson
submitted to Nature in April 1999.

In addition, the following papers are included to assist the newcomer in the field. They will be referred to in the text by their Roman letters.

A *"Microprocessing at the Fingertips"*
Greger Thornell and Stefan Johansson
Journal of Micromechanics and Microengineering, 8, pp. 251-262 (1998).

B *"Lilliputian Reflections"*
Greger Thornell
Proc. of Micro Structure Workshop, Uppsala, Sweden, pp. 24.1-24.7 (1998).

The author's contribution to the papers is as follows:

I	Part of planning. All experimental work and evaluation.
II	Part of planning. Major part of experimental work and evaluation.
III	Part of planning. Most experimental work and evaluation.
IV	Major part of planning, experimental work and evaluation.
V	Part of planning, experimental work and evaluation.
VI	Most planning, experimental work and evaluation.
VII	Part of planning, experimental work and evaluation.
VIII	Most planning, major part of experimental work and evaluation.

CONTENTS

PREFACE

I still remember that morning. With the aim to quickly put my diploma work together, and slush still falling from my shoes I went straight to the room of my younger supervisor, M Sc M. Bexell. Sitting in the chair of his roommate and generating the impatient atmosphere of tapping fingers and frequent consultations of a wrist watch without actually moving a fin, was the head of the micromechanics group, Prof. J.-Å. Schweitz. In a few words he asked me to consider joining the group. His arguments were simple and perhaps not very scientific, but struck me right in the heart. I recall phrases like "it's great fun" and "of course there will be some travelling". Somewhere deep inside I knew I couldn't resist such an offer, and obviously he did too, since he left the room with no apparent reaction to my pretended hesitation. Today I'm pretty convinced Dr Johansson's stomach pulled a few strings in this matter. Not only is this most remarkable intestinal organ of my senior supervisor able to pick suitable recruits to the group (I might not be the best example), but it also announces noon with an accuracy not far from that of the Royal Observatory of London.

In this summary a brief and filtered account of my professional whereabouts since that morning (a period just a little shorter than WW II) will be given. In addition I will try to foist upon you my personal thoughts on micromechanics or microstructure technology (MST), and between the lines, share with you my own reflections on the various projects summarised. This will be made in a simple way, partly because I wish to attract a lot of readers (four is a crowd in this context), but mostly because that's how I like it myself.

1 INTRODUCTION

1.1 THE WUNDERKAMMER LEGACY

Quite on the hype in the Europe of the sixteenth and seventeenth century were carved miniature portraits. The base material for these carvings were fruit-stones and not seldom dozens of, or even a hundred, portraits could be found on a single peach stone. With the invention of the microscope and later the lathe, this fascination for *la sculpture microscopique* grew stronger, especially in Germany where such tiny masterpieces, or *Kunstkammerstücken*, were arranged together with all kinds of exotic material and handicraft into eclectic *Wunderkammern*, or cabinets of wonders. In the Tradescant's Rarities - Essays on the Foundation of the Ashmolean Museum 1683 with a Catalogue of the Surviving Early Collections [1] one can read:

> Other entries refer to items whose particular interest lay in technical virtuosity, such as the 'nest of 52 wooden-cups turned within each other as thin as paper' and various miniature objects: some of the latter were purely technical *tours de force*, such as 'Halfe a Hasle-nut with 70 pieces of housholdstuffe in it', one cherry-stone 'holding 10 dozen of Tortois-shell combs, made by Edward Gibbons' and another 'with a dozen of wooden-spoons in it', a set of Chesse men in pepper-corn turned in Ivory', and 'Flea chains of silver and gold with 300 links a piece and yet but an inch long'.

A simple explanation for this obsession with microminiatures could of course be man's need to control, organise or survey, but one should bear in mind that it evolved in a time of geographical exploration and scientific discoveries. A few decades ago William McLellan, armed with toothpicks and a watchmaker's lathe, replied to a challenge made by the American physicist Richard Feynman, by building a miniature motor no bigger than 1/64th inch cube [2]. Needless to say he rendered a lot of publicity. In 1985 The Gleniffer Press printed "Old King Cole!" on a 22-gms paper and had it bound in a book measuring one by one millimetre - the world's smallest book [3]. In 1990 the Armenian microminiature sculpturer Hagop Sandaldjian [4] passed away leaving a collection of 33 figures most of which had been beautifully carved out of strands of hair carefully selected for texture, tone and tensile strength from the scalps of his family members (people that of course were warded off before the master came to rest by the microscope) and often mounted in the eyes of needles. An exception is shown in figure 1.

Mr Sandaldjian had to hold his breath while carving or painting his speck sized workpieces, but also discovered that his mere heartbeats as transferred through his fingertips (mostly the thumb I guess) were enough to rock his tools and smear or even ruin his work. A single figure could take fourteen months to complete - fourteen months which could be thwarted in a single sneeze. As we shall se later, a sneeze may be most inappropriate also in the technically superior micromechanics workshop of today. With batch processing it could still be a matter of destroying structures having been processed for months, but thanks to modern efficiency a couple of thousand could be destroyed at the same time.

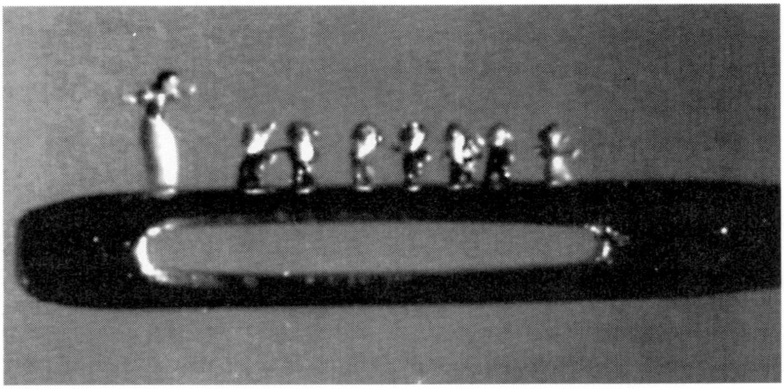

Figure 1. Eight famous characters sculptured out of hair, painted and placed on a needle (courtesy of The Museum of Jurassic Technology).

Without doubt my discovery of an almost complete zoo, carved from ivory and stored in a red bean in one of my father's desk drawers, and a walnut with perhaps 30 or 40 faces engraved on it and sold for a song at a bazaar in Beijing, fertilised my own interest for microstructures, but maybe it was the repeating picture of the label on the "Tomteskur" detergent that really launched my fascination for scaling and miniaturisation. The still flourishing chocolate company Droste uses a similar trick, with a nurse holding a tray with a package of cacao with a nurse holding a tray with a package of cacao, and so on, figure 2. According to the company, the effect has been named the Droste effect and rumours have it that it is even included in Dutch dictionaries.

Figure 2. The Droste effect (courtesy of Droste Holland).

1.2 MICROMECHANICS MATURED

On attending conferences on microstructure technology, MST, one sometimes wonders if it's still just a boyish game. It happens that distinguished, grey suited men just smile and shrug their shoulders when asked why they have fabricated tiny pumps, hover crafts or loudspeakers. And it didn't really help when a 1936 Toyota, complete with license plates, a windscreen and a motor, was fabricated in 1/1000th it's original size [5], although it was just for demonstration purposes and indeed an achievement, figure 3. (Unfortunately, it turned out that it actually ran faster in a sort of vibrating mode with its wheels locked.)

Figure 3. A miniaturised 1936 Toyota (courtesy of DENSO). The surrounding Hindenburg looking things are grains of rice, by the way.

There are however a manifold of applications where MST has entered and proved its strength. My favourite example is the quartz miniature tuning fork of ordinary wrist watches, daily fabricated by the millions [6], figure 4. In cars, air bags are triggered by the deflection of microcantilevers on a sudden speed reduction [7]. Soon micromechanical gyroscopes [8] for skid control will be equally spread. Read/write heads of magnetic hard drives are fine positioned by microactuators [9]. Talking about computers, ink jet printers benefit from small nozzles machined through micro-processes [10]. In scanning probe microscopy, atomically sharp tips are fabricated by similar methods [11].

In contrast to microelectronics, MST hasn't primarily developed for economical reasons or speed potential, but for increased sensitivity and decreased size.

Figure 4. A thimble full of miniature tuning forks for watches.

2
THE SETTING

2.1 A ROAD PAVED BY THE SILICON BANDWAGON

Microstructure technology indeed owes much to microelectronics. A tremendous amount of time and money has been invested in integrated circuit (IC) fabrication during the last 50 years. Silicon, the chief material, has been perfected beyond any other material. Processes for cutting, polishing, doping, cleaning, oxidising, coating, patterning and etching have been developed and optimised to an extent nowhere else to be found. The microprocessor increases its performance one hundred percent approximately every second year. And that's not too bad for a product having had an estimated annual US demand of around 50 pieces, 50 years ago [12].

As the title of this thesis might indicate, my contribution to MST is not very conventional and I would therefor like to bridge the gap by briefly describing some of the steps adopted from silicon based microelectronics. To call them good old processes would be an offence to all those people involved in their important and continuos development, but in some sense they are so spread and considered so reliable that we usually take them for granted.

2.1.1 Through the patterned glass

As soon as the raw material, i.e. the silicon, has been fused together, cut into wafers, polished to a mirrorlike surface smoothness, and cleaned, the unidirectional, planar processing begins. I wish to stress the word planar, because it is indeed the key to efficient and inexpensive manufacturing of electronics, but at the same time rather geometry limiting. One has deliberately chosen a flat substrate and most of the time one subjects it to different kinds of fluxes. It could be a matter of photons, atoms, charged particles or a liquid, but as far as the flux is concerned, there is little need to focus on a particular part of the work piece. (This should be compared to traditional machining where the processing is more localised, no matter if it's a cutting edge or an ablating laser.) To work selectively with some areas of a wafer, the rest has to be protected. This is accomplished by lithography. First a small amount of a liquid, light sensitive polymer, a photoresist, is put in the middle of the wafer. Then the wafer is rotated at a typical speed of one hundred revolutions per second, whereby centrifugal forces spread the substance to a micrometer thin layer on top of the wafer. The solvents responsible for the relatively low viscosity of the photoresist are removed in a soft baking step in a hot air oven or on a hot plate and the wafer is ready for exposure. Prior to this, the desired pattern has been designed by CAD and drawn by a laser on a template - a relatively thick glass plate with a thin chromium layer on it. Through the openings in the chromium layer of this template, the wafer is exposed. Then, depending on the nature of the resist, the exposed areas are either made more stable by cross linking the polymer, or degenerated by chain scissoring (negative and positive photoresist respectively). In subsequent development (usually by immersion in an alkalic liquid) the less stable parts of this thin masking layer are dissolved preferentially and the wafer has been patterned, figure 5.

Before the resist can be used for anything worth mentioning it has to be treated with heat once again (hard baked) to be even more stabilised, or cross linked. After this a number of things can be done. Through the openings, material could be deposited by different methods either by a low energy process, like evaporation, to give a thin layer

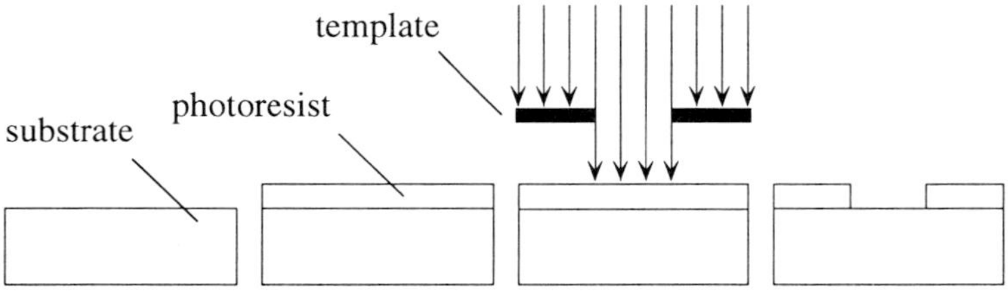

Figure 5. Lithography in cross section, with deposition of photoresist, exposure through template and developing (the photoresist is assumed positive).

of material on top of the wafer, or in a more brutal way to bury, or implant, atoms deeper into the material, figure 6. The first type is typical for final metallisations of electrodes, while the second, or doping as it's called, is used for intermediate steps in IC fabrication.

Figure 6. The adding of material through the openings in the masking material, either on top of the substrate (as with evaporation), left, or inside it (as with ion implantation), right.

One can also subtract material locally through the openings by etching. By sticking to low temperature processes (say below 130°C) the photoresist can still be removed afterwards in ordinary solvents like acetone, and the cycle repeated as many times as necessary as long as the flatness of the wafer is preserved to some extent. (In case of material deposited on top of the photoresist it will peel off together with the resist during "stripping" by a process called lift-off.) For microprocessors it could easily amount to 20 cycles, while one or two cycles are usually enough in micromechanics. For high temperature processes, like diffusion doping or some deposition methods, or aggressive etchings, the photoresist doesn't work very well as a masking material. Instead the wafer is first either coated by for instance a metal layer or much more often, corroded at high temperature in oxygen to form an oxide layer at the surface. Lithography is then used to pattern this layer which in turn could act as a stable masking material for processing the underlying material.

This is the basics of micromachining and I believe Intel's Gordon Moore alluded to a whole bunch of ancient Greek philosophers, when he used the words: earth, wind, fire and water to describe the fabrication of a microprocessor, being a question of melting together and oxidising one of earth's most abundant materials, and then repeatedly rinsing it in extremely pure water.

2.1.2 On nature's premises

On etching the wafer, at least two things could happen. If the material looks the same in all directions, the consumption of it will generally proceed at equal rates in all directions. If on the other hand, the atoms have arranged themselves periodically into a so called single crystal (figure 7), etching will proceed at different pace in different directions. (Standing in the middle of a major parking lot and trying different bearings one will register different periodicities in different directions. In a three-dimensional atomic lattice one speaks of crystallographic planes of different packing densities. And, just like a parked car being more easily liberated in some directions, the difficulty of removing an atom depends on the angle of attack relative to it's dearest neighbours.) These behaviours, resulting in rounded pits or trenches or faceted holes or grooves, are referred to as isotropic and anisotropic etching respectively, figure 8. The former is typical for disordered materials like amorphous glass or fine grained, or polycrystalline, materials like most metals, while the latter occurs for highly ordered materials, such as the semiconductors used for electronics.

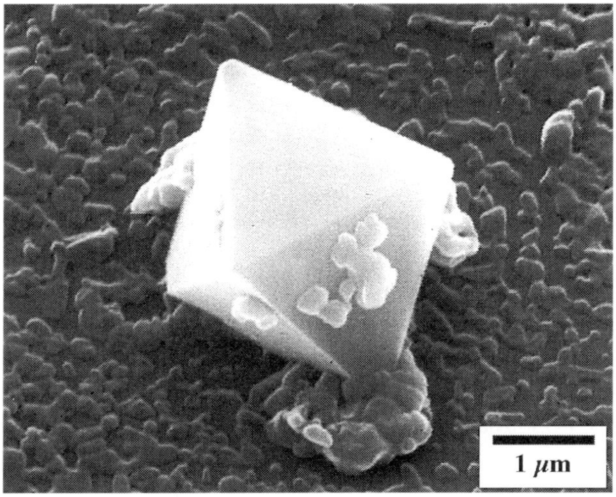

Figure 7. A single crystal found on an etched, but poorly rinsed, quartz surface. (In synthetisation of silicon the same perfection is found throughout metre sized ingots.)

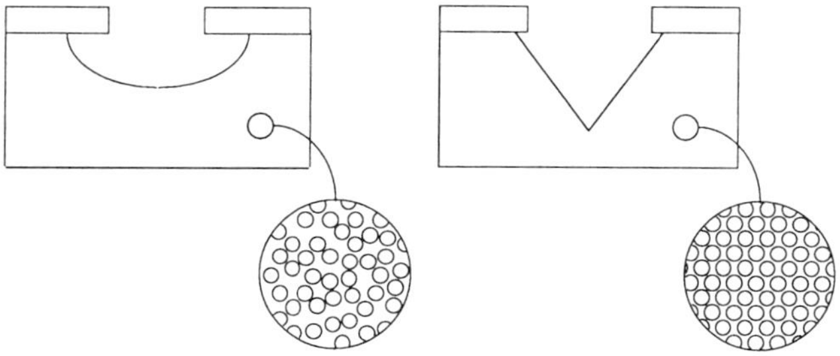

Figure 8. Cross sectional views of isotropically and anisotropically etched trenches or pits in disordered and ordered materials respectively.

Also in ordered materials it is possible to get isotropic looking etch profiles by choosing a mass transport limited (and thus agitation dependant, figure 9) etching instead of the reaction limited process assumed above.

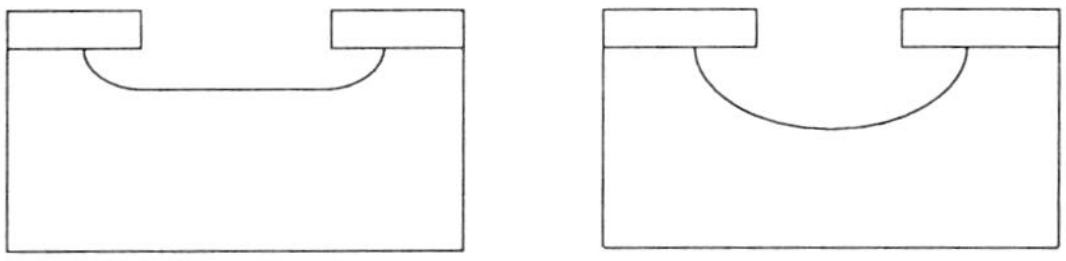

Figure 9. The effect of agitation on depth and uniformity of isotropic etching (left: poor, right: good).

Historically, controlled dissolution of silicon seems to have started in the late fifties with isotropic etching [13]. In the mid sixties anisotropic etching is mentioned as a way of isolating components in a more space efficient way [14]. For the electrical engineer the atomic order of the material allows him to predict electrical properties, while it gives the micromechanic a way of accomplishing corners, nice edges and flat surfaces.

Ways of circumventing the sometimes limiting anisotropy are dealt with in some of the papers of this thesis [II-V].

2.1.3 Surface vs. bulk

Traditional micromachining (especially that of silicon) could be divided into two types depending on the role of the wafer. If it is used mainly has a substrate or carrier for material deposited on top of it, one speaks of surface micromachining [15], but if it's shaped and actually used in the final structure, one talks about bulk micromachining [16]. For a given structure the dominant method could usually be determined at a quick glance. Generally, surface micromachining is characterised by thin (a few micrometers), stacked layers with intricate lateral shapes, while bulk micromachining results in thicker and often homogeneous structures (tens or hundreds of micrometers), often with a geometry in harmony with the single crystalline material. Although some intermediate layers (sacrificial layers) are often etched away to facilitate local release of gears and levers in surface micromachining, the sandwichlike result in cross section bears a resemblance to integrated circuits. Other than by its size, the result from bulk micromachining is difficult to distinguish from that of an ordinary workshop.

2.1.4 Dry vs. wet

As most people probably associate etching with copper engravings or decoration of glass, I temporarily omitted the important adjective "wet". Dry etching is not performed in a simple beaker on a hot plate, or in a wet bench. And in contrast to dry cleaning, it's not even close to a liquid. In fact it is carried out in a place a lot drier than any living creature has ever experienced - a plasma. In addition, a plasma is a rather busy place [17]. Charged particles are accelerated in a low pressure gas with the help of an applied electric field, they bump into each other, or far more likely, into neutral atoms and molecules, get excited, emit photons and glow. Compared to the less ordered diffusion or convection governed mass transport of etching species in a liquid, often assigning much authority to the reactivity of the workpiece material, the mass transport in a plasma could be made much more directional. In the physical regime resulting in a somewhat anisotropic looking result, charged particles and low pressure are used. By this a particle, typically a positively charged ion, will follow the electric field (perpendicular to all surfaces) with little chance of getting scattered by other particles, and atomically blast, or sputter as the experts say, the substrate on hitting it. This so called ion beam etching (IBE) gives rather steep side walls but unfortunately consumes the masking material almost as quickly as the underlying material. If the pressure is increased from the say 10^{-4} torr (a pressure less than one millionth of the air we breath) common for IBE, charged particles will bump into other particles more often and loose energy as well as sense of direction on their way to a surface. In its least ordered form, at a pressure of maybe 1 torr, the plasma based etching benefits from the dissociation or excitation of gases to more reactive species. One example is the ashing of organic materials (e.g. photoresist) in an oxygen plasma. Here one works in the chemical regime with a much higher materials selectivity, but looses the directionality and gets an isotropic etching, figure 10. This kind of dry etching is simply referred to as plasma etching (PE).

Figure 10. An example from plasma etching of silicon (the neck of this snowman structure is about 5 μm in diameter).

With a pressure somewhere in between (10^{-3} -10^{-2} torr) we have the popular and compromising reactive ion etching (RIE) relaying not on sputtering, but on the mere energy contribution from impinging particles. Contrary to sputtering with its quite complex dependence of yield on angle, full advantage can here be taken of the directionality of the ion bombardment and the aspect ratio can be increased even more. When for instance silicon is etched in chlorine, the reaction products require ion bombardment to be desorbed. With the adding of hydrogen to CF_4 on etching silicon, a similar effect is obtained, but due to a simultaneous reaction. Carbon starts to polymerise rapidly at all surfaces and only those constantly cleared by ion bombardment are etched. Bringing it one step further, material could be redeposited on vertical surfaces to increase the aspect ratio even more.

This brief account for dry etching is illustrated schematically in figure 11.

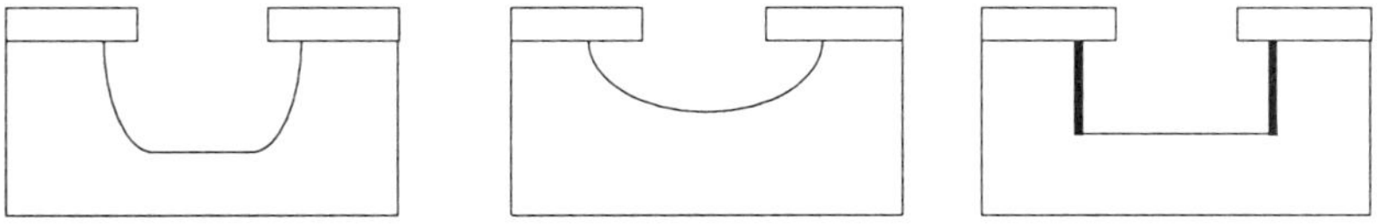

Figure 11. The various regimes in dry etching; ion enhanced etching or sputtering (left), plasma etching with low ion contribution (middle), and ion enhanced etching together with passivation of vertical surfaces (right).

Currently, two methods exist to exploit this passivating (or inhibiting) concept further. One is to cryogenically cool the substrate to stop spontaneous etching on side walls, and the other to constantly switch between an etching and polymerization step using different gases and voltages [15].

Since one of the thesis's paper [VI] deals with sputtering, it should be said that sputtering is also the name of a deposition technique where instead of the substrate being negatively biased to attract ions, the material to be deposited is made cathode, or target as it is usually called. Atoms sputtered away from the target are then caught on the substrate to form a thin film. Compared with evaporation, which is a less expensive thin film deposition technique where atoms are condensed from the vapour of a lump or wire of material heated to melting at a far lower pressure, sputtering yields a film of relatively uniform thickness, allows for many more materials to be deposited and provides some means of controlling the energy supplied to the growing film through adjustment of pressure and acceleration voltage.

2.1.5 Figures of merit

There are a few concepts frequently used in micromachining to compare one process to another. The first is resolution, almost always referring to the narrowest line attainable. Of course a high resolution is preferred in most cases. Selectivity is the difference in consumption rate between two materials subjected to the same subtractive process. In most lithography based etching, it is important to have high selectivity between the thin masking material and the substrate to reach some depth. Underetching is the widening of for instance a trench under the edge of the masking material. Limiting the final resolution, underetching should be kept to a minimum. Aspect ratio finally, is the depth (or height) to width ratio of a structure. A high aspect ratio is usually impressing.

Having said this, I can safely direct you to paper A for a further discussion on microstructuring. In fact I encourage you, since it is a vital part of this summary, although the brief discussion on conventional microprocessing above should be a background solid enough for the following treatise. To justify the thinking behind some of my contributions to this field, I would next like to describe the arena of conventional micromachining, an environment I do not dare to calculate the number of hours I have been in.

2.2 THE WALK OF THE BUNNY SUIT

Given that the biggest threat to successful microfabrication is the human operator and supervisor, chip manufacturer would most certainly prefer a fully automated factory where silicon gravel is shovelled in at one end and packaged devices are shovelled out by the other end. That's because even the tiniest speck of dust entering a process would be equally good an obstacle for IC fabrication as a petrol truck right across a highway, and humans are really good at generating particles. (Around half a million particles fall off a human being standing still - each minute.) The solution to this dilemma is isolation. First of all the manufacturing premises, or the *clean room*, are protected from outdoor contaminants by sophisticated filters and a minimum of air inlet. Secondly, all materials used in the "fab" must generate a minimum of particles (no matter if it's the mask aligner or a wet bench, the material of the floor or the paint of the wall, or a pen or a tissue). Finally the operators are dressed up in dust proof, hooded overalls, visors, gloves, and shoes. In addition they are told to shave regularly, move slowly, just speak if necessary and never sneeze in the *clean room* areas. Despite their buggy dressing, covered faces and silence, you can still identify your colleagues by their height and, most of all, by their walk. Add to the cleanliness, a yellow light and a controlled humidity (not to excite the photoresist), a devibrated floor (not to disturb sensitive alignment equipment and electron microscopes), a perforated floor and ceiling to facilitate laminar air flow, constantly blinking diodes, and humming and buzzing equipment calling for your attention, and you can soon visualise one of mankind's strangest working places. When it all works you can reach a pretty high class cleanliness, or a low class *clean room*. Usually the performance of a *clean room* is measured by the number of particles per cubic foot of air exceeding say 0.5 µm in diameter. A "class hundred room" is sufficient for most microfabrication, while a "class 1,000 000 room" is better used as an office. At home, sand may constitute a problem in ordinary vacuum cleaning unlike most other visible particles we wish to get rid off. In the *clean room* the lighter particles are the real enemy. To remove barely visible dust (size of 50 µm) from a surface by 50%, an airflow of about 30 m/s is necessary, while an equally efficient elimination of 15 micrometer particles requires 100 m/s [18].

Part of my investigations treats non-*clean room* alternatives [I,VII,VIII], and in paper B scaling phenomena, like that of gravitational and surface forces acting on objects of various size, are discussed.

2.3 THINGS PUT INTO PERSPECTIVE

From the section above, it's obvious that size matters. Microelectronics are far ahead of micromechanics in reduced size. A thumb-nail-sized Pentium chip has around six million transistors on it. Its interconnections are submicron in width, and if these were enlarged to the width of an ordinary office corridor, the chip would cover 100 square miles, an area larger than greater Stockholm.

In general, micromechanical structures seldom measure less than 100 micrometers by their largest dimension. Most structures have overall dimensions of millimetres and are thus clearly visible by the naked eye. Often however, one dimension is severely suppressed. It could for instance be the thickness of a pressure sensitive membrane or the cantilever of the earlier mentioned airbag sensor. And by suppressed, I mean down to one micron.

As a comparison the diameter of a human hair is between 50 and 100 micrometers (μm) depending on age, race and sex. The resolution of the eye is somewhere in the vicinity of 30 μm. A red blood cell is seven to eight micrometers in diameter. A period of visible light is about half a micron long, while 5,000 to 10,000 atoms in a row equal 1 μm. Finally, ordinary black ants are around 4,000 μm long (i.e. 4 mm), figure 12.

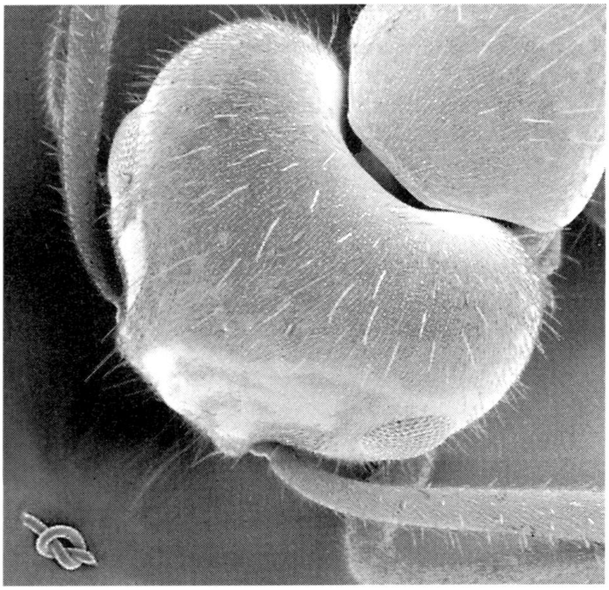

Figure 12. A well-known dried and gold/palladium sputtered creature photographed with the help of a scanning electron microscope.

2.4 NO LONGER A CHIP OF THE OLD BLOCK

As said in paper A, it might be that micromechanics evolved from microelectronics. And I won't argue with you if you claim most micromechanical devices are still being fabricated by the conventional methods outlined above, but there's a lot more to it.

For at least ten years, a constant stream of new and innovative microprocesses has threatened to drown us. Many of them lack a macroscopic counterpart, but one can discern an effort to include traditional machining and stretch the limits. Most of these balder attempts reach us from Japan while Europe and the US seem to stick to the good old IC technique to a larger extent.

When R. P. Feynman had promised the USD 1,000 reward to the one that could build a motor fitting inside a 1/64 inch edge cube (i.e. just about covering the period ending this sentence), and Mr McLellan later came to collect it, he felt a little robbed. This, partly because he was newly married, had some loan pay off to do, and completely had forgotten to tell his wife, but mostly because he had hoped to see the development of some new technologies. Today, I can come up with a least two non-traditional ways of fabricating the coils for McLellan's electromagnetic motor (he himself, simply wound them from a 10 μm diameter wire). One is the laser assisted CVD [19] mentioned in paper A, and the other is the localised electrodeposition described in paper VII.

The reason for developing new technologies are mainly material or geometry related. So far cost and speed appear subordinate. But why then is IC-technology still prevailing? I'm not sure, but I can see one reason (and it's not very convincing) for not exchanging one USD 100 000 equipment for another, or for batch manufacturing 1,000 components in three hours instead of serially processing one component in 10 s, when the gain in geometrical or material freedom is positive. That single reason could be compatibility. A true benefit of flat, IC derived processing is the possibility to align subsequent process steps quickly and accurately to previous ones.

At the moment the true reason doesn't really matter. From a processor's point of view, the important thing is that we are able to process, on a minute scale, materials as odd as diamond, concrete, DNA and fullerenes (figure 13), or work truly three dimensionally with certain polymers and metals, almost as if drawing a structure in free space (see paper A and VII). In fact the tool kit is so complete, it begins to be a matter of finding real problems to good solutions (but that was the case also for the laser and microprocessor, and we all know what happened). In conclusion microstructuring still clings to processing used for microelectronics, but branches fast and often.

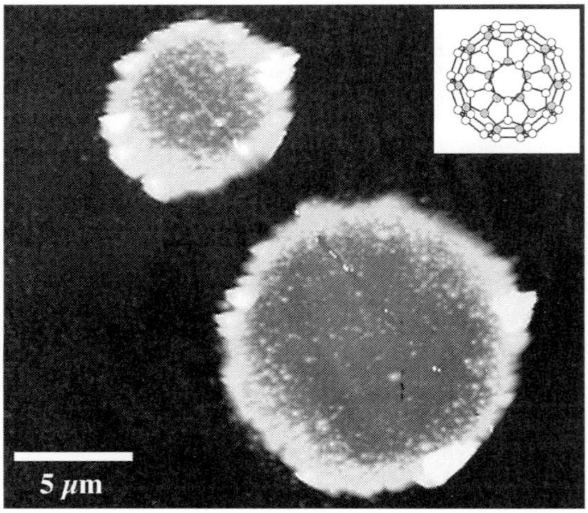

Figure 13. I know, this could be almost anything, but believe me, it is C_{60}-molecules, or buckyballs, polymerised on platinum (the darker areas). The insert is a drawing of a C_{60}-molecule.

2.5 MY EFFORT

At least twice I have touched upon my own mission. Part of it consists of finding ways to arbitrarily shape materials that would either show great anisotropy or isotropy in conventional microprocessing. This is achieved by adding or subtracting material voxelwise (paper I, VII and VIII), or by superimposing or even inducing an anisotropy in unruly, single crystalline, or amorphous materials respectively (paper II-V). The second goal is to investigate alternatives to clean room bound processing, both because of the enormous investment connected to building one, but also because of the quite unexpected cost of running one. The solution to this is to work quickly in small confined volumes with inexpensive equipment (paper I and VII-VIII). I like to refer to this as desktop microprocessing.

In the paper summaries to follow, I do not intend to sell these techniques to you. On the contrary, I will try to point out weaknesses, difficulties and disadvantages, how difficult it may be for a proud parent.

3 PAPERS CONDENSED

3.1 SUMMARY OF PAPER I

Among its workshop mates, it indeed is a slow method to shape materials - the technique employed in paper I: "Design and Fabrication of a Gripping Tool for Micromanipulation", but getting tired of etching and mounting tungsten needles to tweezers for our micromanipulator, we decided to give it a try.

In wire electro-discharge machining (WEDM) the tool, a thin wire, and the workpiece are made electrodes in an RC-circuit. Both are submerged in a dielectric liquid and brought close together. Suddenly the electric field strength will cause the liquid to break down into a very localised plasma which will heat, melt, and on ceasing, tear off a minute drop of material from the workpiece. The tool is moved a little, a new spark appears, and by this a contour could be traced much like in band sawing. On a larger scale the geometry is governed by the relative movement of the electrodes in two dimensions, and by the inclination of the wire to the workpiece surface in the third. The resolution is given by the positioning accuracy, while the minimum slot width of course is determined by the diameter of the wire.

Our design for a miniature gripper was flat and for special reasons we had a titanium alloy in mind. Was it not for a combination of external actuation transferred through pliable elements, lack of space, and the need for a jaw stroke of one millimetre, WEDM alone would have been a perfect process. Unfortunately the resulting surface skin, a heat affected and brittle layer, is prone to cracking and not at all suitable for the kind of strains intended here. By subsequent etching we removed this layer and greatly improved the topography, which by itself should decrease stress concentrations.

Finite element analysis was used to determine the length of the pliable elements and levers to facilitate full closing of the jaws safely within the capability of a piezo-actuator fitting into the system without exceeding the yield stress of the structure at any point, even with a hypothetical maximum load and some misalignment.

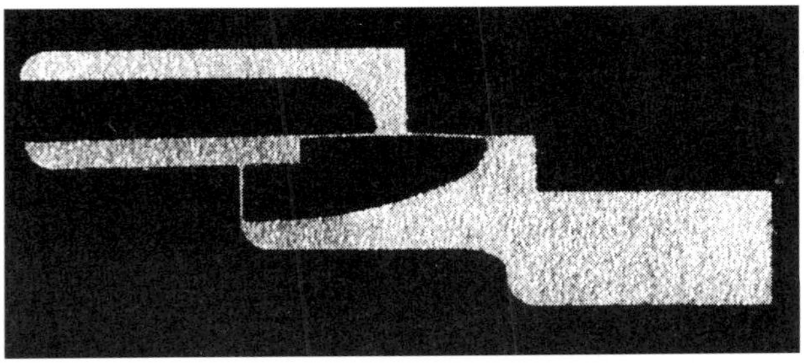

Figure 14. A gripping tool fabricated by electro-discharge machining followed by etching.

The finished tool is given in figure 14 and a close-up of the central part after etching, is shown in figure 15. When the upper jaw is deflected, the lower jaw is subjected to a small torque which makes it responsible for closing about 15% of the gap.

Experiments, where wires of different dimensions and materials were liberated from the jaws to establish the gripping force, indicated a linear relation of force to elongation of piezo driver, and naturally a dependence on the coefficient of friction.

Figure 15. The improved surface finish with wet etching.

This work was the natural sequel to my diploma work which was more process than device oriented. It also gave me an excuse to study the gripping mechanisms of insects, much later inspiring me to paper B.

3.2 SUMMARY OF PAPER II

In the paper "Heavy Ion Induced Etch Anisotropy in Single Crystalline Quartz" [II], we investigated the possibility to induce, or rather superimpose, a virtual anisotropy by bombarding quartz wafers with swift, heavy ions prior to conventional lithography and etching. This, to satisfy man's desire for square and circular construction geometries (probably stemming from the need to treat designs analytically) also in highly anisotropic materials.

On passing through a solid, a really fast (e.g. 10,000 km/s) ion seems little aware of the target's nuclei, instead it interacts preferentially with its electrons. If the solid happens to be insulating, there is a temporary shortage of electrons which makes the nuclei undergo coulomb repulsion resulting in atomic defects. Such defects are extremely thin (typically 100 Å), very straight, long (tens or hundreds of microns), and denominated latent ion tracks. If subjected to etching the disordered track will dissolve much faster than the surrounding, perfect crystal. This is pretty much it - at low magnification, the result from high selectivity looks like that of high anisotropy.

What we did was to stochastically, but unidirectionally, bombard single crystalline quartz wafers of different crystallographic orientation (Z- and AT-cuts), coated with a thin gold/chromium masking material, with different doses of Xe-ions (around 10^8-10^{10} ions/cm^2) and with two different angles of incidence (90° and 45°) with the help of an accelerator. The masking material was then photlithographically patterned and the substrates etched in a hot alkali solution.

In bare areas the tracks were developed into pores which, on further etching, widened and finally merged, figure 16, leaving vertical and inclined structures of lateral geometries otherwise difficult, or even impossible, to achieve in quartz, figure 17. In addition we confirmed the possibility to heal the undeveloped tracks at high temperature (1200°C), but realised that this high a temperature was not beneficial for the material's piezoelectrical properties. A somewhat surprising result was a decrease in edge and corner definition with higher ion doses. We ascribed this to the higher probability of encountering dense clusters of tracks with higher fluences.

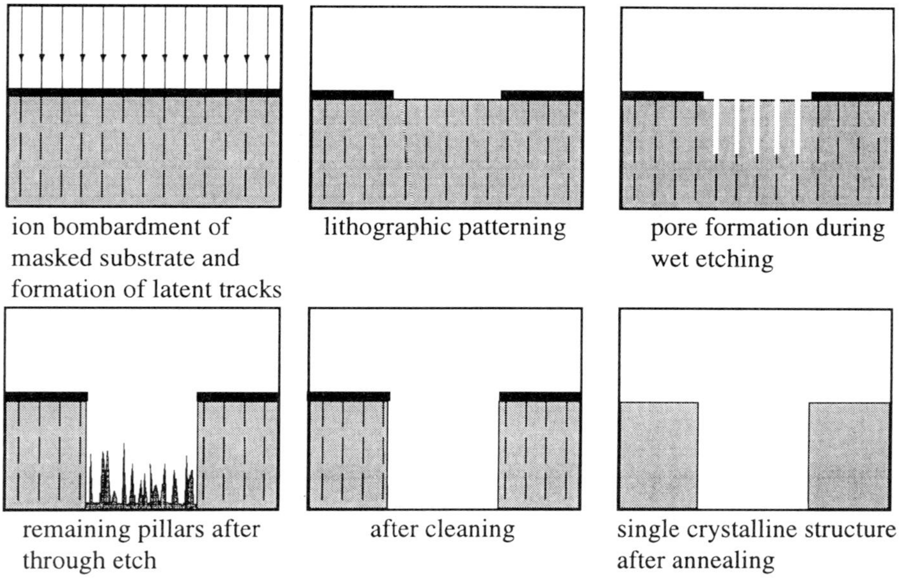

Figure 16. Micromachining by ion track etching in schematics. Broken lines represent latent ion tracks.

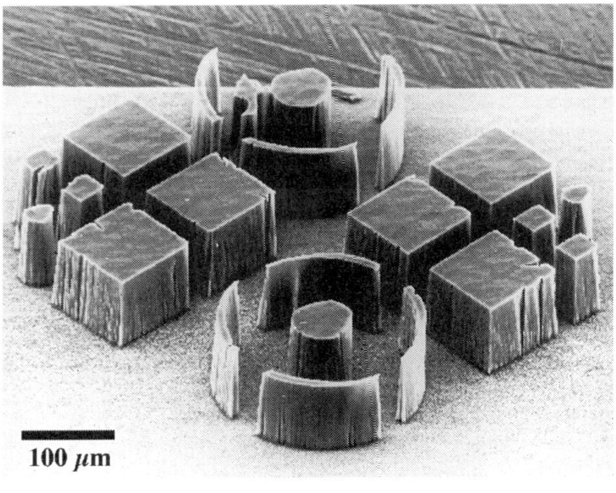

Figure 17. Test structure etched in AT-cut quartz.

We baptised the method MITE (Micromachining by Ion Track Etching), mainly to annoy those people exploiting ants to prove resolution and aspect ratio of their processing techniques. This far though, we haven't been able to recruit an obliging mite for our PR campaign.

3.3 SUMMARY OF PAPER III

As the title "Anisotropy-Independent Through Micromachining of Quartz Resonators by Ion Track Etching" implies, this was a follow up of the objective outlined in paper II. Together with a Swiss manufacturer of miniature quartz tuning forks for wrist watches, we investigated the feasibility of MITE as an alternative to conventional wet etching, to improve structure symmetry.

To cope with the intended structure thickness (125 µm) and the allotted accelerator energy, we used double sided ion bombardment. To comply with industrial efficiency, we tried different etchants (13 M KOH at 135°C, 15 M NaOH at 125°C and $2HF:3NH_4F$ at 55°C, by themselves or combined) together with different ion fluences (10^9 and 10^{10} ions /cm^2), and equally important, we irradiated commercial fork resonators and measured their quality (in terms of Q-value), and temperature stability to see if latent tracks in this piezoelectric material was of any harm.

A lot was learnt from this. With ion tracks overlapping from the two sides of the wafer, a zone sandwiched by two straggling layers (the very last percent of the ions' range where nuclei interaction starts) formed. When etched, whole pieces of material were removed from this layer giving a coarse appearance, figure 18. The idea of a two-step etching (with KOH as a very track selective etchant for opening, and a second faster etchant for widening) was advantageous only for the high fluence samples, but not for speed reasons. The NaOH etchant was found to severely attack the adhesion layer of chromium and then the underlying ion tracks through pinholes in the gold layer. The top dog for low dose samples was the buffered hydrofluoric acid which, with an exception for the sandwiched layer, gave good looking structures at short time. (Fortunately these are also chemicals used in conventional quartz etching.) From the irradiated resonators, one could spot no degeneration with the lowest dose ($2 \cdot 10^8$ /cm^2) while a two orders of magnitude higher dose gave a decrease in Q-value of around 10%, and parabolic frequency vs. temperature dependence shifted to a higher temperature, figure 19. With such promising results we estimated the cost for including irradiation as an preparatory step in tuning fork fabrication and reached the quite astonishing conclusion that it would be a matter of a few percent of the total cost (USD 0.25) for a trimmed and vacuum sealed component.

On the whole, the double sided irradiation left a lot to be desired, but at least we found a solution to the problem with webbed tines, and got an idea for a different ion track based process technology [IV].

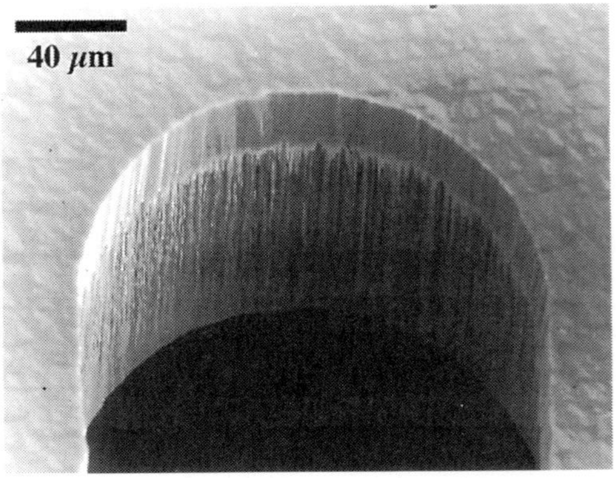

Figure 18. The result from double sided irradiation followed by etching, showing the problem with the "sandwiched layer".

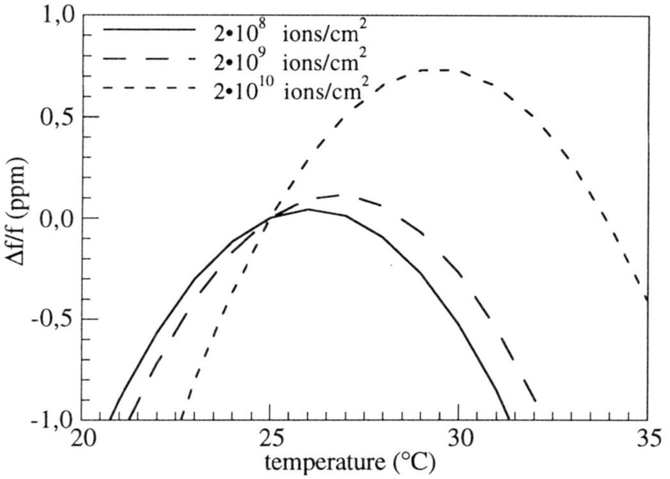

Figure 19. The frequency vs. temperature dependence for commercial tuning forks irradiated with $2{\bullet}10^8$, $2{\bullet}10^9$, and $2{\bullet}10^{10}$ ions/cm^2. The first curve is not possible to distinguish from that of an unirradiated resonator.

- So, there is actually a layer of ions covering the bottom of the structures?
- Eh! Well, at the higher fluence, there should be just one ion per million substrate atoms. And besides, it is probably removed during etching.

(Being misunderstood at a conference.)

3.4 SUMMARY OF PAPER IV

With paper IV: "Micromachining by Ion Track Lithography", we temporarily abandoned the idea of tailoring the anisotropy of substrates for general purposes, and instead decided to define entire latent structures by bombarding the substrate with heavy ions through a mechanical mask (most often a thick LIGA fabricated stencil mask), figure 20 (left).

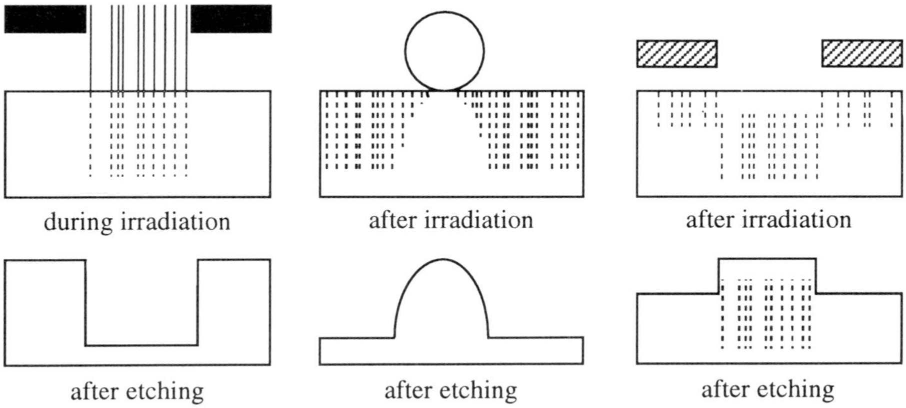

Figure 20. Ordinary ion track lithography employing a stencil mask (left), grey scale scheme with a thickness modulated mask (middle), and negative scheme with a semi-transparent mask (right).

The advantage was at least twofold. First we didn't have to worry about latent tracks in the remaining structure as before, and secondly with a high enough selectivity or dose, the need for a mask during etching became obsolete. We practised this triumphantly on a number of materials like quartz, mica, glass, polycarbonate and polyimide (see figure 21 for the last one). In contrast to ordinary photolithography it even turned out to work perfectly well on already structured substrates. The major findings were a significant improvement of edges with higher ion doses, because of the shorter etch time necessary, and a rare but nasty scattering of ions under the mask, which in absence of an etch mask of course resulted in misplaced pores.

Thanks to the small spread in range of the ions, grey scale lithography, figure 20 (middle), was possible and we successfully projected a strand of hair into quartz.

Then there is that second phenomena of an ion sometimes having to speed down a little before it can cause etchable damage, which we used to demonstrate negative lithography, Fig. 20 (right). The part of the beam blocked with ordinary "scotch magic" gave etchable tracks all the way to the surface, while tracks in other areas were buried beyond reach of the etchant.

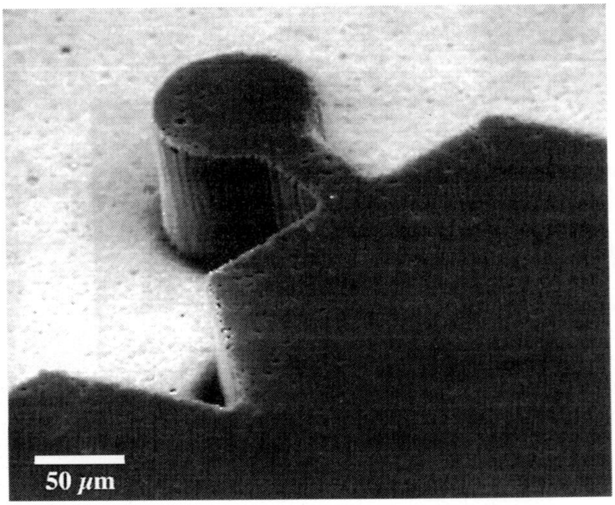

Figure 21. A test structure made in polyimide.

When sitting by all the monitors and Star Trek-like panels in the control room of a particle accelerator in the middle of the night, the words of Leo Szilard sometimes come to my mind:

We turned the switch, saw the flashes, watched for ten minutes, then switched everything off and went home.

3.5 SUMMARY OF PAPER V

At this time we wished to use the ion track lithography for something useful, and aimed at fabricating a miniature tuning fork resonator from x-cut instead of z-cut quartz. In conventional tuning fork fabrication, one benefits from the prismatic planes of the Z-cut to define tines of rectangular cross section. Unfortunately, the piezoelectric properties (i.e. the phenomenon of strain, and hence deflection, induced on applying an electric field, used for activation of the vibrations), are not matched with the etch anisotropy of the material, and a certain sacrifice has to be made in field uniformity. In addition the tines have to be metallized on all four sides which is a little tricky. With ion track lithography we have the ability to define rectangular beams in any cut, and can really show off in the X-cut, which lacks natural, orthogonal planes, but is characterised by a strong piezoelectric coupling. By this, electrodes are needed on only two sides of the structure, figure 22.

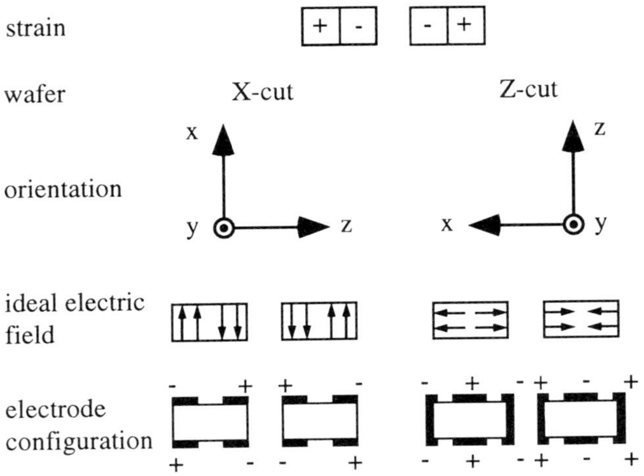

Figure 22. The two tines of a tuning fork in cross section with the strain, ideal electric field and actual electrode configuration for x- and z-cut devices.

In contrast to ion tracks in the Z-cut [II], which opens up rather symmetrically, the X-cut contains the z-axis, which is the fastest etching direction in conventional wet etching. To find the most suitable etchant for this cut, we conducted a short pore opening study, figure 23, which made us reject the hydrofluoric acid and instead use the sodium hydroxide, which in the absence of masking material no longer posed a threat (cf. [III]).

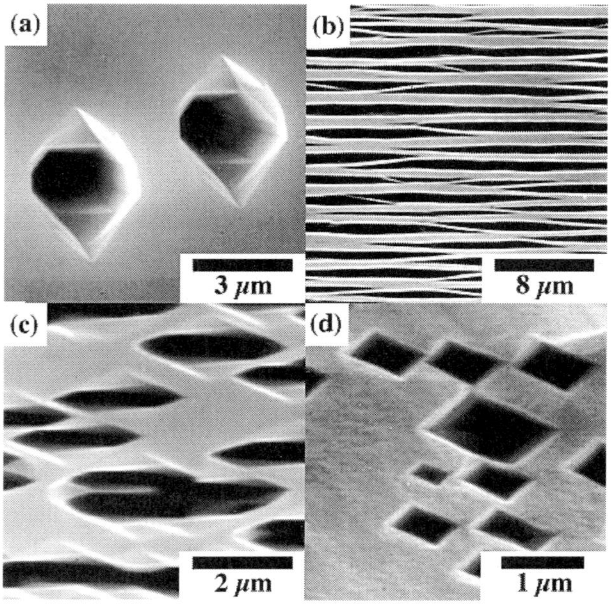

Figure 23. Typical pore geometries resulting from different etchings of equivalent ion tracks. From upper left and clockwise: NaOH, HF, LiOH, and KOH. The z-axis is horisontal in all panels.

A 40 kHz tuning fork realised by this method is shown in figure 24.

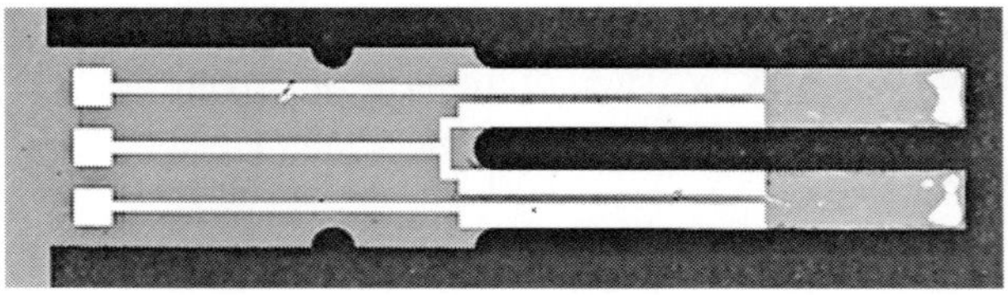

Figure 24. The finished x-cut tuning fork (length of tines approx. 2.5 mm).

A real challenge with this scheme is the metallization and electrode patterning. It is always difficult to apply planar processes to non-planar structures. In our case we got metallic and polymeric residues at the structures (see figure 24) which we believe lowered the Q-value (here being 42,000 or less, as compared to 50,000-80,000 for commercial miniature tuning forks). Of course we could use a combination of conventional and ion track based lithography to solve this, but then we would have to find another etchant, and include an extra process step, making the technique less attractive.

At resonance, a vibrating body very efficiently exchanges energy with its exciter. This manifests itself by large amplitude. For a simple experiment, tie five or six rubber bands together, fasten a magazine (or this book) to one end, and let the whole assembly hang freely from your hand by the other end. Then, if you slowly rise and lower your hand, the magazine will follow with the same amplitude. At too high a frequency the magazine will appear at rest while the rubber bands do the work. At a frequency between these extremes the response amplitude of the magazine will be greater than the driving amplitude.

3.6 SUMMARY OF PAPER VI

In paper VI wet etched micromechanical cantilever beams (figure 25 right) were used to monitor the internal stress of sputter deposited quartz resonator gold electrodes, the reason being that the stress affects the apparent stiffness and hence the resonance frequency of the component.

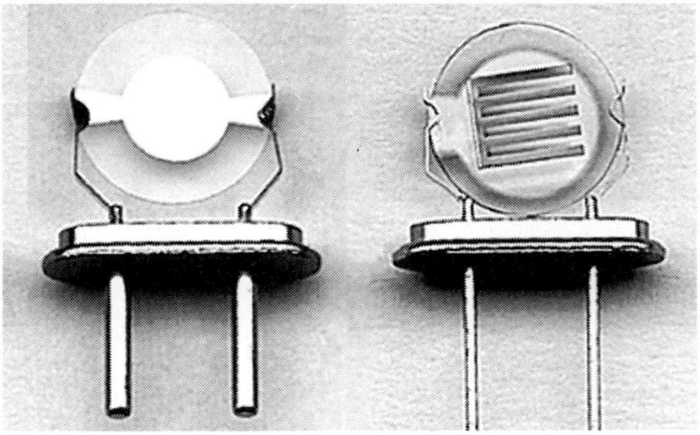

Figure 25. An ordinary 13.2 MHz resonator (left) and a stress monitor sample (right) mounted in the same way. (Beams are 3.8 mm long and typically 10 μm thick.)

Approximately 10 μm thick cantilever beams were etched out of quartz by standard lithography and hydrofluoric etching. These beams were coated with gold at different process conditions (pressures, deposition rates and temperature) and the resulting deflections were translated into film stress. Measurements were conducted repeatedly for a period of 17 weeks to investigate the relaxation of stress with time. After four of these weeks, the ageing was accelerated by raising the temperature from room temperature to 85°C.

For all samples the resulting initial stress was relatively low and compressive (ranging from -182 to -63 MPa), but turned tensile for a few samples with time and annealing. Given that all substances involved were inert, that the contents of argon (as determined with Rutherford backscattering spectrometry) was below one atomic percent, and that the absolute stress increased with lower pressure, the stress was blamed on atomic peening*.

By means of transmission electron microscopy, the grain structure was found to be much coarser at lower deposition rate. Also, it was found to grow coarser with time and annealing except in the case where the deposition and annealing temperature were similar.

Since the stress itself poses little problem for the frequency stability other than by acting as the driving force for relaxation, attention was paid not only to absolute stress levels, but to the relative change with time. It was found that a sample coated at 85 °C exhibited a much smaller decrease in stress with time compared with others of both higher and lower initial stress.

On calculating the expected frequency shift from the stress encountered with the parameters representing commonly deposited electrodes, a shift of 4 ppm for a 13.2 MHz AT resonator was obtained, which is in good agreement with observations from real resonators.

---

*This is the same stress/strain generating mechanism as that utilised to finely adjust the curvature of some aircraft parts by shot peening.

In *On Electrostriction* [sic], Dr E. J. Mills observed a certain "rise in temperature" on plating thermometer bulbs, and was actually able to measure internal stress in degrees centigrade [20].

3.7 SUMMARY OF PAPER VII

The idea of localising electrodeposition by using a very fine electrode, as described in paper VII is relatively new and potentially important in microstructure technology since it facilitates the building of high aspect ratio metal structures of intriguing geometry. Basically, one puts a needle electrode close to the surface of a negatively biased substrate immersed in, or at least covered with, a plating solution, figure 26, and then withdraws it as metal ions from the electrolyte are reduced to metal.

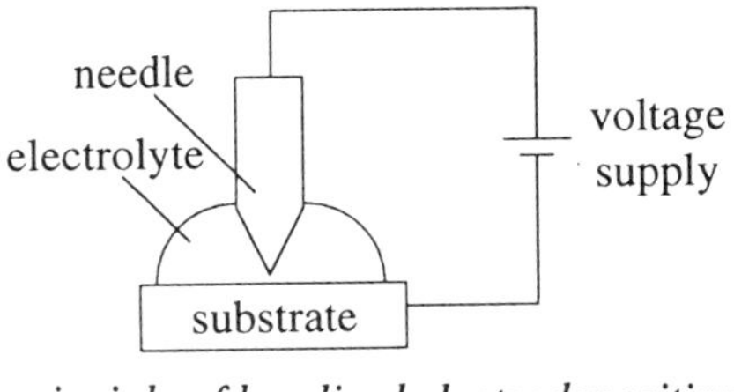

Figure 26. The principle of localised electrodeposition in schematics.

In order to improve resolution and surface smoothness, we found it necessary to investigate the effect from adding complexing agents and supporting electrolytes to aqueous solutions of various nickel salts. In particular, solutions of nickel sulphamate and nickel sulphate with ammonium formate and ammonia were studied. In addition we tried to minimize the stray deposition on nearby and previously built microstructures, by using a third electrode - a working electrode. Finally, we took advantage of this unlocalised deposition to broaden and join two separately built structures.

Both structure width and surface roughness decreased with increasing amount of ammonium formate in nickel sulphate based electrolytes. At a concentration of 3.0 M and voltage of 4.2 V, structure diameters of 2-3 µm with almost mirrorlike finish were accomplished. On increasing the voltage to 4.8 V, structure widths decreased even more, but at the expense of surface smoothness. The severe unlocalised deposition experienced with nickel sulphamate was suppressed when the working electrode was connected to the substrate.

In figure 27, structures illustrating the potential of the technique are given.

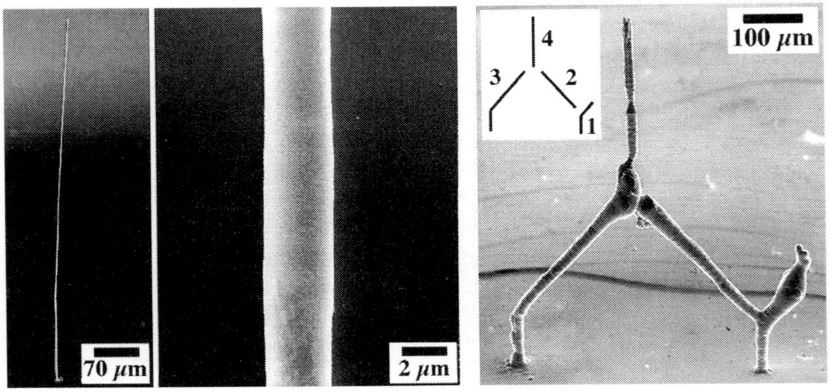

Figure 27. A 600 µm tall structure with an aspect ratio of 200 (left and middle) and the result from joining two structures (right). (The insert in the right panel shows the building sequence.)

A disadvantage with some of the electrolytes tested was poor adhesion in combination with cracking, making it difficult to build structures higher than approx. 7-800 µm.

On looking for these thin structures on their relatively large substrates in the microscope, an appealing feeling of aviation appeared:

.. and there, far below, a little offset from the center, is a column of lights pointing upwards, changing angles as I fly - the Eiffel Tower.

Charles Lindbergh

3.8 SUMMARY OF PAPER VIII

In the paper "Desktop Microfabrication - Initial Experiments with a Piezoceramic", we used a home built droplet dispensing unit with an anisotropically wet etched nozzle, figure 28, to deposit a slurry of PZT particles, basically an organic vehicle with small (one micron) well dispersed particles of the piezoceramic material PZT. PZT is a material of growing importance in microstructure technology because of its large piezoelectric effect (strains of around one per mill), but also representative of a whole class of materials currently being difficult to micromachine.

With the help of a heatable substrate facilitating fast evaporation of the solvent, we were able to turn from formless blobs built at RT and 40°C to millimetre high structures with typical diameters in the range of 200-300 µm at 60 and 80°C, by depositing droplets on droplets at frequencies of 25 or 50 Hz for typically 30 or 40 seconds.

Figure 28. A silicon nozzle with a newly ejected droplet (diameter 40 μm).

On closer investigation the structures were found to be hollow, figure 29, with typical wall thicknesses of 10-20 μm. From observations of the building process, we assume the early formation of a vapour bubble being the reason for this hollowness. At the highest temperature tested, 80°C, the end of the structure was observed to inflate and

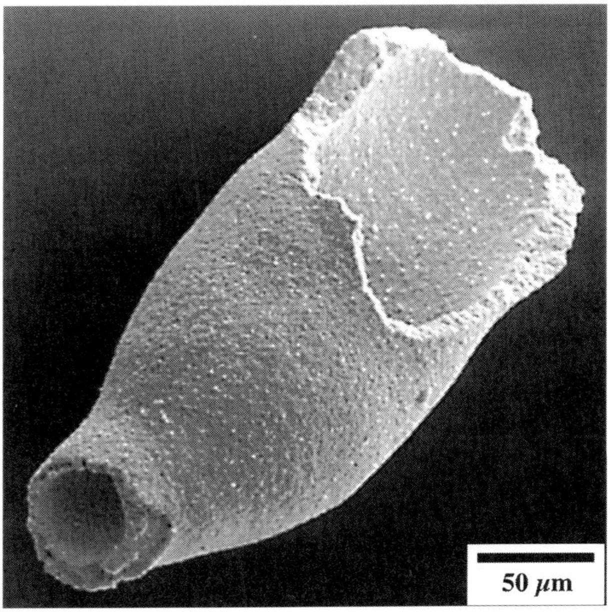

Figure 29. A drop-by-drop fabricated and sintered PZT structure.

collapse repeatedly during building up, which might explain the sagging of some structures. Furthermore we found a steady state, liquid and semi-transparent meniscus on the tip of the growing structure, consisting of approximately 50 droplets, and withdrawing within a second from termination of the building process. Because of this meniscus, ring shaped segments found on the green bodies and corresponding in number to the droplets, appear circumferencially even if the individual droplets hit the structure off-axis.

We concluded that the dispensing unit worked best with low viscosity slurries and that promoted drying was essential for building high aspect ratio structures. Unfortunately the higher drying temperatures and lower frequencies beneficial for the structures tend to result in clogging of the nozzle.

In parallel we ran a few "kitchen experiments" involving coffee, hot plates, medical carbon and soap solutions, not at all to entertain ourselves, but for the sheer adding to our understanding of particle transport and stain formation.

4
WRAPPING IT UP

4.1 SHAPELESS SHAPING

If you are still with me and haven't neglected the references (especially paper A), I guess you have realised how diffuse the boundaries of micromachining are. During my years in this field I have come to the conclusion that it is seldom the resulting size, but more often the origin of the process that makes one speak of micromachining. Take the watch tuning fork for instance. If defined lithographically and wet etched, as it most often is, it's clearly a micromachined device, and it doesn't matter that it is a few millimetres long. Now, in figure 30, there is a 25 kHz langasite ($La_3Ga_5SiO_{14}$) tuning fork fabricated by cutting, and (at least as long as the lithographically patterned electrodes are disregarded) most micromechanics would say it has been high precision machined, despite the size being the same. The same holds for the Sandaldjian sculptures (figure 1) in comparison with the LIGA-fabricated gear in figure 31, but then of course the resolution could be questioned.

It's a shame by the way that we don't reach the same accuracy in MST as we do in high precision machining. In the former we often settle with a relative accuracy of one per cent or one per mill, while we have parts-per-million-tolerances in the latter.

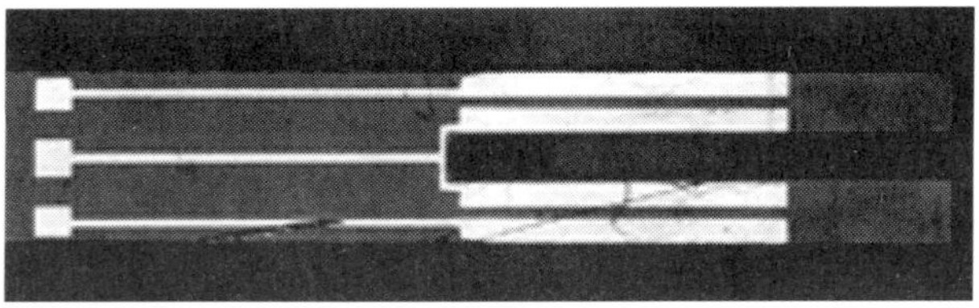

Figure 30. A langasite miniature tuning fork fabricated by cutting. (Note the webbed tines due to the curvature of the blade.)

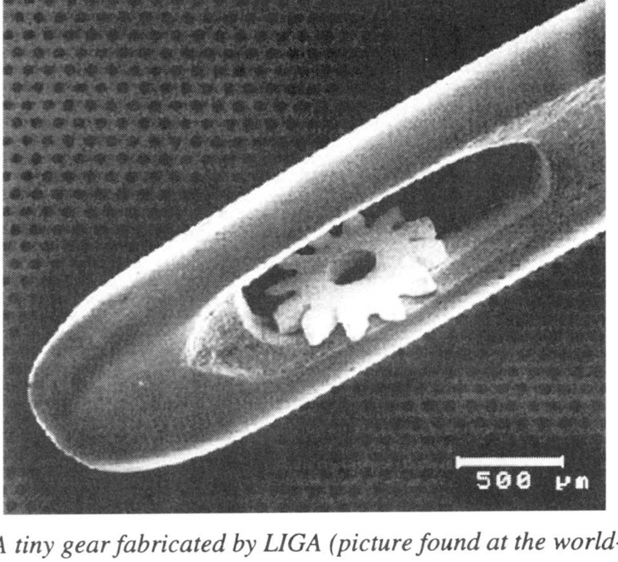

Figure 31. A tiny gear fabricated by LIGA (picture found at the world-wide web).

I can see a trend towards incorporation of high precision machining in MST [A], and maybe, just maybe, we then have to predate MST to the time of fruit-stone carvings.

4.2 "DON'T TELL ME ABOUT MICROFILM"

In the Middle Ages, scholars discussed how many angels would fit on the head of a pin. Although some engraving was done to find an answer, it was mostly a philosophical question. In his famous talk "There's plenty of room at the bottom" in 1959 [21], Richard Feynman discussed to what degree the Encyclopedia Britannica (approx. 25,000 pages) had to be demagnified to fit on a head of a pin. A factor of 25,000 is the answer. Still, that means the diameter of a dot in the half-tone reproductions would be the size of 1,000 atoms. Bringing it one step further, he proposes writing it all in code with each letter represented by six or seven bits, each being a small cube of around 100 atoms, and taking full advantage of the material, i.e. using also the volume and not only the surface. By this one should be able to squeeze in all the information of all books in the whole world, in a speck of material barely visible by the naked eye. Feynman ended his talk by offering a prize to the one who could reduce a page of text by 25,000 times (linear scale). Fortunately, at least for his marriage, it took 26 years before Tom Newman at Stanford with the help of an electron beam wrote Charles Dicken's "A tale of Two Cities" at this scale and claimed the prize [22]. (The main technical difficulty seems to have been to locate the text once it was written. Even the head of a pin becomes an huge area at that magnification. This I recognise from my own experiences from electron microscopy.)

Today, when single atoms are mechanically arranged into text or pictures (albeit at cryogenic temperatures) in scanning probe microscopes, one wonders what Feynman would have said. But do we really need this minuteness?

4.3 FUTURITY

In MEMS (Micro Electro Mechanical Systems) we won't need fabrication techniques with atomic resolution. In fact we very seldom need to reduce the whole system to the microdomain. We need sensitive, and thus minute, mechanical structures like the cantilever beams in collision sensors. The microprocessor handling the signal has to be small for speed reasons and economy. But for the moment we don't have to make the whole system small. In the car industry they still use ordinary bolts to attach components. And the enclosure must withstand a mechanic leaning on it while reaching for a dropped washer in the engine compartment. The all-on-one-chip-idea will be necessary the day we wish to "swallow our surgeon", but not much earlier.

Here, it feels appropriate to emphasise the necessity of learning from nature. Being such macroscopic, and statistically out-evening objects, as we human beings are, we find it difficult to understand the surface forces acting on a tiny silicon beam or sympathise with a water molecule being beaten roughly 10^{15} times a second by its fellows (at RT). In our field, insects serves as illustrative prototypes. They don't rely on pneumatic or hydraulic systems for mass transport. And evolution has already eliminated insects with a taste for showers and baths. We don't have to go so far as to decapitate cockroaches and command them by small virtual reality helmets, or connect stimulating electrodes to their legs as Kuwana et al. [23], but we already have, and should, benefit from for instance the little resonating systems facilitating the rapid wing beat of flying insects [24], or the gyroscopes behind the skilful navigation of ordinary house-flies [25].

Before the birth of MST, Feynman anticipated surface micromachining, and various kinds of beam technology, but also proposed another way of building tiny machines [21]. It was a kind of *top-down*-approach where conventional machines, like lathes, were building similar machines and tools one fourth their size, which in turn build even smaller machines and so forth *ad diminutandum* (cf. the Droste effect). In 1981 Eric Drexler put forward the opposite [26] - a *bottom-up*-approach, the goo concept, or molecular engineering as I suppose he likes to put it himself. He envisioned the synthetisation of almost anything atom-by-atom or molecule-by-molecule in a beaker or a test tube, much like the assembling of protein chains.

Other than as metaphors, I personally don't believe in these ideas. As said in paper A, there is often a need to change the method of processing when decreasing size. I suppose surface forces make it difficult to get rid of micrometer sized lathe chips and shavings. On the other hand, when this debris reach nanometer size it would perhaps be impossible to hold on to, at least in a Brownian environment. In a broader sense though, I agree with Feynman. In paper VIII, we used a micromechanical device to build microstructures, but mark well, we changed the technology completely. Drexler's vision I share in an even more remote way. To get maximum geometrical freedom, I believe in point-by-point processing (see [VII]), but I'm not convinced of atomic assembling being fast enough to generate micrometer or millimetre sized structures in reasonable time (at nanometer level it's a completely different thing). Or am I? I have witnessed the precipitation of rather large and perfect crystals within minutes from saturated electrolytes and etchants.

I doubt it makes me more of a materials scientist saying it, but on putting an imaginary cross in the periodic chart for each element I have come across during my Ph D studies, it reminds me of a far advanced game of battleship. To throw in a substance with possible trace elements this close to the end won't hurt, especially since it illustrates a promising trend in microsculpturing. The concrete mentioned earlier was structured by embossing or moulding. Standard compact discs with their micrometre sized depressions are made by replication. This kind of replication benefits from the extremely fast (matter of seconds) and accurate filling of a master with a flowing, or at least ductile material. I really believe in this technique despite the obvious negligence in paper A.

As a sort of tribute to one of earth's most appreciated substances [27] and a tying together of this summary I would like to show the great material flexibility of this method, figure 32. Manufacturing the silicon master by lithography and plasma etching took a few hours, but the "tepid" embossing less than a minute.

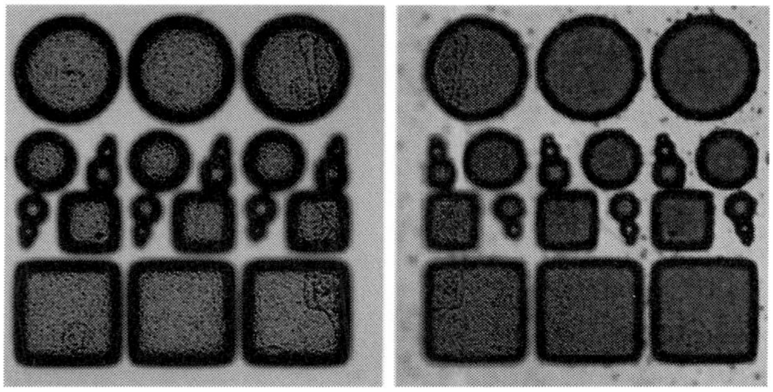

Figure 32. A tasteful microstructure (right) made by replication of a dry etched silicon master (left). (Total length of bar approx. 0.5 mm.)

ACKNOWLEDGEMENTS

To the following persons, institutions, constellations etc., I owe much. Without them I couldn't have thrived in my selfish business to the extent I did.

Thank you:

- my supreme commander: Jan-Åke Schweitz for talking me into this, for your relaxed attitude to science, and for all those special assignments embellishing and extending my working days,
- my supervisors: Stefan Johansson, Klas Hjort and Fredric Ericson for all helpful and inspiring discussions, support in the laboratories, for letting me do what I wanted, and for arranging all boring (but creative I suppose) financing behind my back,
- the division's two mothers: Carin Palm and Ann-Sofie Lindberg for taking care of all the administration, for nursing the whole division, and for being such good friends,
- the division's two fathers: Rein Kalm and Jan-Åke Gustafsson, for indefatigably instructing me on all the sensitive and fragile equipment, for helping me build experimental apparatus, and for repairing all wrecked things (and Rein, 17 metres - that's not much!),
- my dear co-workers: Mats Bexell, Henrik Bengtsson, Håkan Rapp, Lena Klintberg, Anders Jansson and all the rest for excellent work and friendship,
- the entire Materials Science Division for creating such a friendly atmosphere (had it been more friendly, I wouldn't have done any experiments or writing at all),
- the people at the Solid State Electronics Division - in particular Christer Hedlund and Pelle Rangsten, for teaching me fluttering, sputtering and lithography, for sharing with me, their knowledge in quartz processing and electroplating, and their good mood,
- Guy Portnoff and Jan Söderkvist for kind, rapid and external assistance,
- the instrument makers for turning, drilling and milling so well from my vague but impatient instructions,
- the librarians at NTC for fast delivery of all kind of reading material (and Monica, I really appreciate that express faxed reference ([23]) I got an hour ago),
- The Swedish Foundation for Strategic Research (SSF), The Swedish Research Council for Engineering Science (TFR) and The Swedish Board for Industrial and Technical Development (NUTEK) for financial support,
- my other "sweet piece of cake" - Eva for all support and love, for listening to my tiresome and daily accounts for setbacks and successes, and for catching that drowsy ant for me (figure 12), together with Margot.

Uppsala, April 1999

'Ford!' he said, 'there's an infinite number of monkeys outside who want to talk to us about this script for *Hamlet* they've worked out.'

Douglas Adams, "The Hitch Hiker's Guide to the Galaxy"

REFERENCES

[1] *Tradescant's Rarities Essays on the Foundation of the Ashmolean Museum 1683 with a Catalogue of the Surviving Early Collections*, edited by Arthur MacGregor, Clarendon Press, Oxford, 1983.

[2] *No Ordinary Genius - The Illustrated Richard Feynman*, edited by Cristopher Sykes, Weidenfeld & Nicolson, London, 1994.

[3] Guiness Rekordbok, Forum, 1991.

[4] *The Eye of the Needle - The Unique World of Microminiatures of Hagop Sandaldjian*, Society for the Diffusion of Useful Information Press, West Covina, 1996.

[5] M. Hisanaga, T. Kurahashi, M. Kodera, and T. Hattori, *Fabrication of a 4.8 Millimeter Long Microcar*, Proc. of the Second International Symposium on Micro Machine and Human Science, Nagoya, Japan, pp. 43-46(1991).

[6] E. Momosaki, *A Brief Review of Progress in Quartz Tuning Fork Resonators*, Proc. of IEEE International Frequency Control Symposium, Orlando, USA, pp. 552-565(1997).

[7] G. T. A. Kovacs, *Micromachined Transducers - Sourcebook*, WCB McGraw-Hill, New York, 1998.

[8] J. Söderkvist, *Micromachined Gyroscopes*, Sensors and Actuators A, 43, pp. 65-71(1994).

[9] D. A. Horsley, M. B. Cohn, A. Singh, R. Horowitz and A. P. Pisano, *Design and Fabrication of an Angular Microactuator for Magnetic Disk Drives,* J. of Microelectromechanical Systems, vol. 7, no. 2, pp. 141-148(1998).

[10] A. Asai, K. Shinjo and H. Miyake, *Present and Future of Ink Jet Printing*, Proc. of the International Conference on MIPE, Tokyo, Japan, pp. 575-580(1997).

[11] B. W. Chui, *Microcantilevers for Atomic Force Microscope Data Storage*, Kluwer Academic Publishers, 1999.

[12] M. S. Malone, *The Microprocessor - A Biography*, Springer-Verlag, New York, 1995

[13] H. Robbins and B. Schwartz, *Chemical Etching of Silicon I. The System HF,* HNO_3*, and* H_2O, J. of the Electrochem. Soc., vol. 106, no. 6, pp. 505-508(1959).

H. Robbins and B. Schwartz, *Chemical Etching of Silicon II. The System HF,* HNO_3*,* H_2O*, and* $HC_2H_3O_2$, J. of the Electrochem. Soc., vol. 107, no. 2, pp. 108-111(1960).

[14] J. H. Forster and J. B. Singleton, *Beam-Lead Sealed-Junction Integrated Circuits*, Bell Laboratories Record, 44, pp. 313-317(1966).

a less secretive work might be:

D. L. Kendall, *On Etching Very Narrow Grooves in Silicon*, Appl. Phys. Lett. vol. 26, no. 4, pp. 195-198(1975).

[15] J. M. Bustillo, R. T. Howe and R. S. Muller, *Surface Micromachining for Microelectromechanical Systems*, Proceedings of the IEEE, vol. 86, no. 8, pp. 1552-1574(1998).

[16] G. T. A. Kovacs, N. I. Maluf and K. Petersen, *Bulk Micromachining of Silicon*, Proceedings of the IEEE, vol. 86, no. 8, pp. 1536-1551(1998).

[17] B. Chapman, *Glow Discharge Processes*, John Wiley & Sons, New York, 1980.

[18] L. Månsson, *Högteknologins osynliga fiende*, Timab Tvätt i Malmö, 1992

[19] K. Williams, J. Maxwell, K. Larsson and M. Boman, *Freeform Fabrication of Functional Microsolenoids, Electromagnets and Helical Springs Using High-Pressure Laser Chemical Vapor Deposition*, Proc. of IEEE MEMS, Orlando, USA, pp. 232-237(1999).

[20] E. J. Mills, *On Electrostriction*, Proc. of the Royal Society of London, 26, pp. 504-512(1877).

[21] R. P. Feynman, *There's Plenty of Room at the Bottom*, J. of Microelectromechanical Systems, vol. 1, no. 1, pp. 60-66(1992). (Transcript of talk given the annual meeting of the American Physical Society in 1959.)

[22] L. Weschler, *Mr. Wilson's Cabinet of Wonder*, Vintage Books, New York, 1995.

[23] Y. Kuwana, N. Watanabe, I. Shimoyama and H. Miura, *Behavior Control of Insects by Artificial Electrical Stimulation*, Distributed Autonomous Systems, Springer Verlag, Tokyo, pp. 291-302(1994).

[24] R. McNeill Alexander, *Exploring Biomechanics - Animals in Motion*, Scientific American Library, New York, 1992.

[25] *Control of Equilibrium in the Flying Insect*, Nature, vol. 163, pp. 74-75(1949).

[26] K. E. Drexler, *Molecular Engineering: An Approach to the Development of General Capabilities for Molecular Manipulation*, Proc. Natl. Acad. Sci. USA, vol. 78, no. 9, pp. 5275-5278 (1981).

[27] By now I begin to wonder if talking to myself isn't the most noble form of "private communication". According to me chocolate is extremely popular.